Billie the Brain Learns to READ

Written by Dr. John Hutton

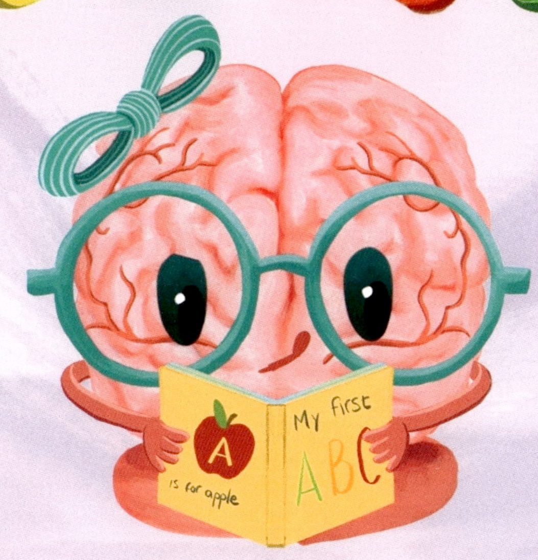

Illustrated by Gabriella Vagnoli

Editorial Design by Mayte Suarez

Published by blue manatee press, Highland Village, Texas.
blue manatee press and associated logo
are registered trademarks of Arete Ventures, LLC.

First Edition: September 2025.

Library of Congress Cataloging-In-Publication Data
Billie the Brain Learns to Read / by Dr. John Hutton;
Illustrated by Gabriella Vagnoli–1st Ed.
Summary: Learn how the brain is "wired" for reading, guided our superstar friend, Billie the Brain! See the benefits of families reading together, from cuddly feelings and bonding in infancy, to language development, to learning letters, letter and word sounds, imagination and meaning. The main text is simple, rhyming and fun, accompanied by scientifically accurate brain illustrations and facts about reading milestones. Written by a pediatrician and expert in early literacy and brain development, this is a perfect book for families, educators, advocates and children to share and learn!
ISBN-13 (paperback): 979-8-9886382-9-2
[1. Juvenile Fiction - Concepts/Body. 2. Juvenile Fiction–Health and Daily Living.]
Printed in the USA.

The text for this book was set in Adobe InDesign.
Artwork was created digitally.

To my mom,
who read fantastic books to me
and took me to the library;
to my daughters,
who inspired me to follow her example;
and to Shelley,
who always reminds
me of how fortunate I am.
- JSH

To my sister Paola,
who taught me how to read
and who I'd be lost without.
- GV

Billie the Brain
was relaxing in her **cranium**
one sleepy night,
snuggly feelings flowing
in the glow of a bedside light.

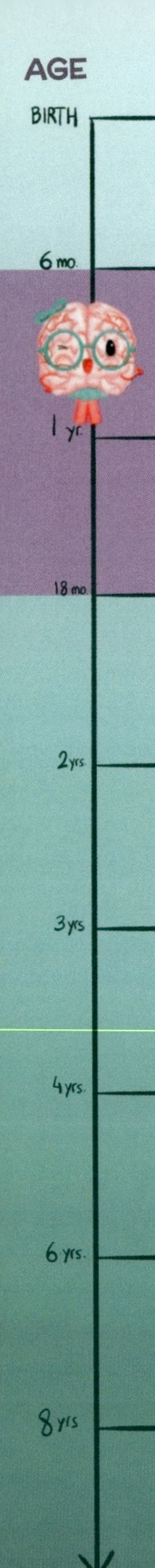

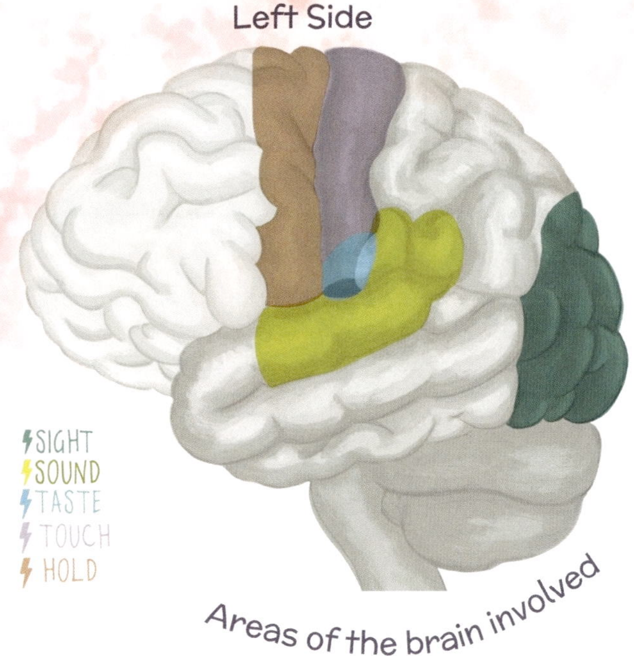

It's normal for babies to *taste* books!

Early Print Concepts:
Babies first learn what books are and how they work using their *senses.*

Loving hands showed Billie a *thing,*
inviting her to explore.
It was **hard**, smooth, **not so tasty**
- with colors, faces and more!

AGE

BIRTH

6 mo.

1 yr.

18 mo

2 yrs

3 yrs

4 yrs

6 yrs

8 yrs

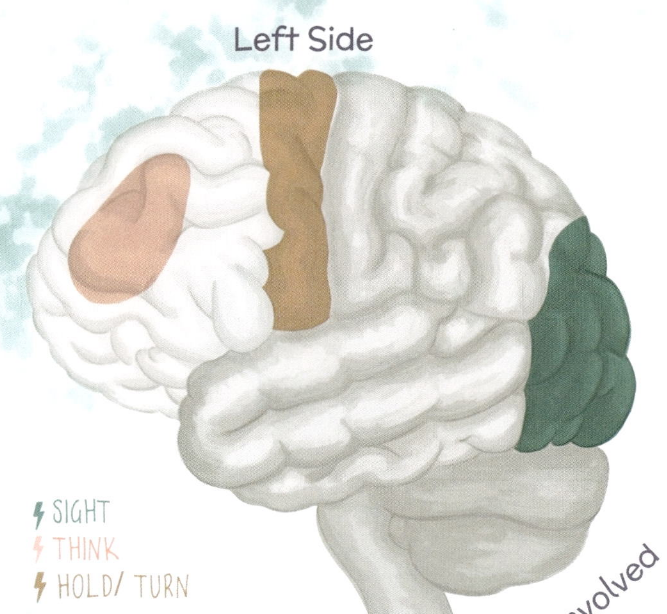

Later Print Concepts:

More advanced parts of books.

Pictures and words tell **stories**.

Squiggly marks on pages are **writing!**

Stories have a

beginning,

and **end.**

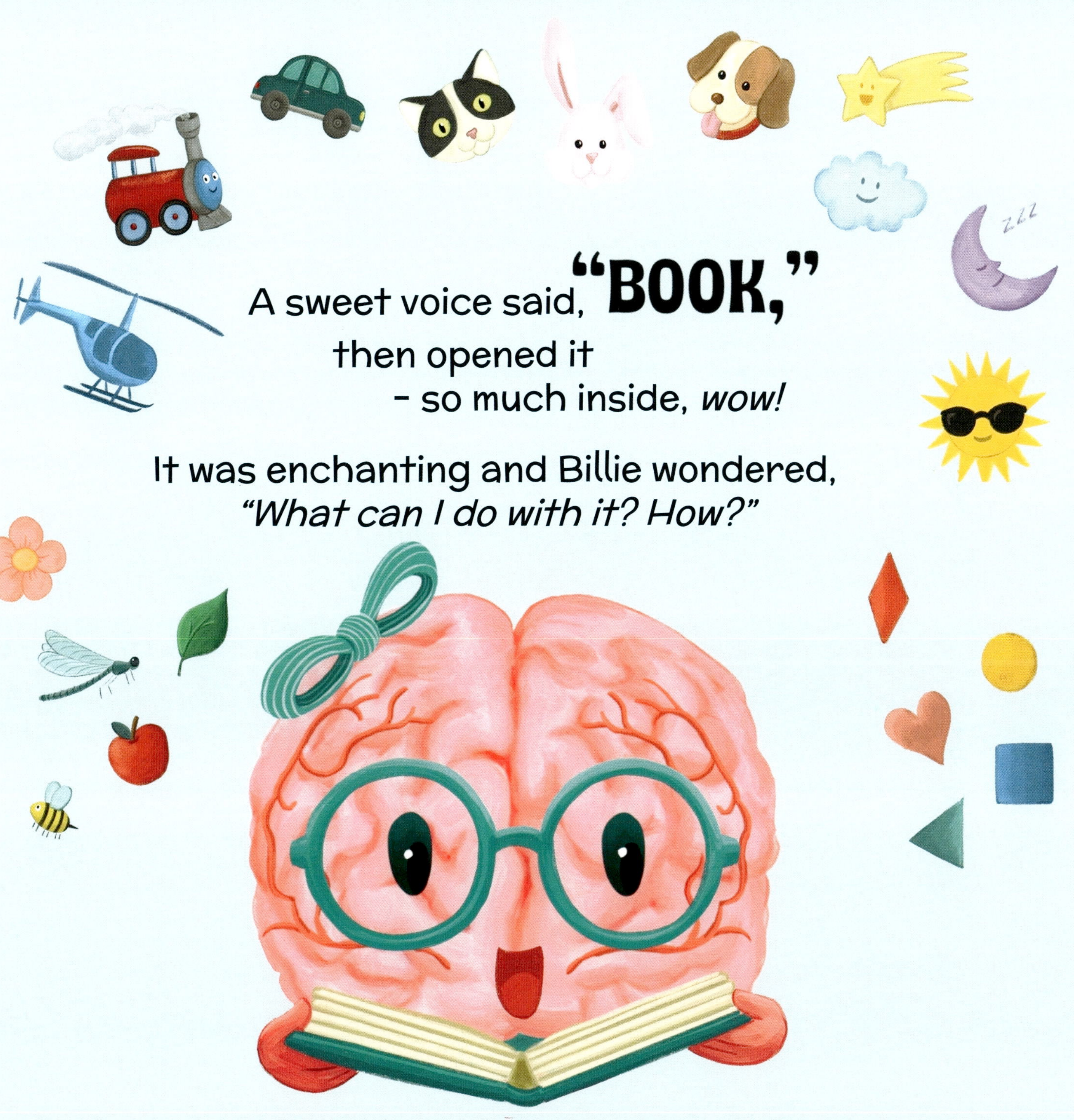

A sweet voice said, **"BOOK,"**
then opened it
- so much inside, *wow!*

It was enchanting and Billie wondered,
"What can I do with it? How?"

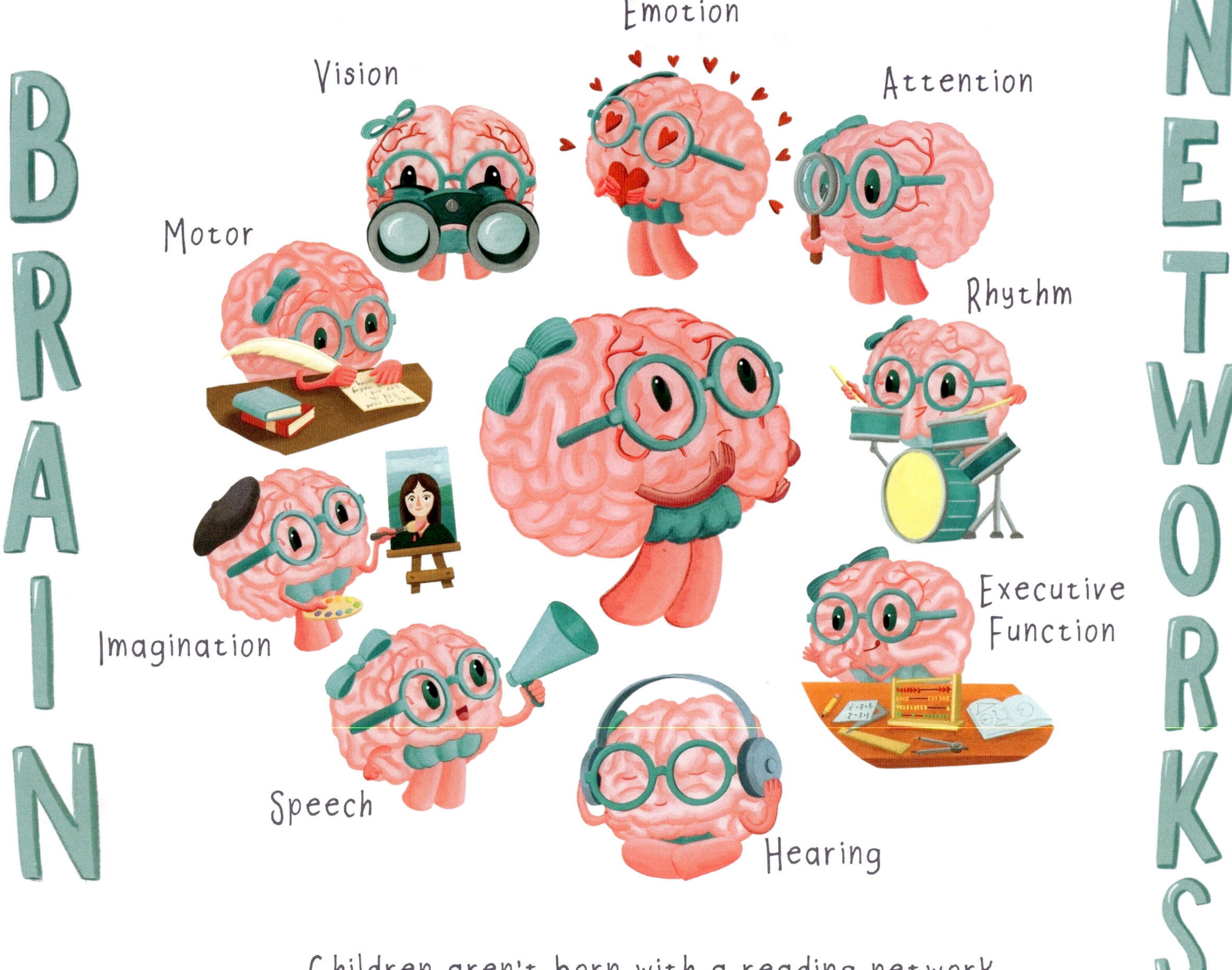

Children aren't born with a reading network.
They have to build one through ***practice.***
This involves connecting brain networks for skills that they are born with, especially **vision** and **language** (speech and hearing).

Her *a-Billie-ties* were amazing
but Billie did not know how to read.
A new **network** built from ones she had
was exactly what she'd need.

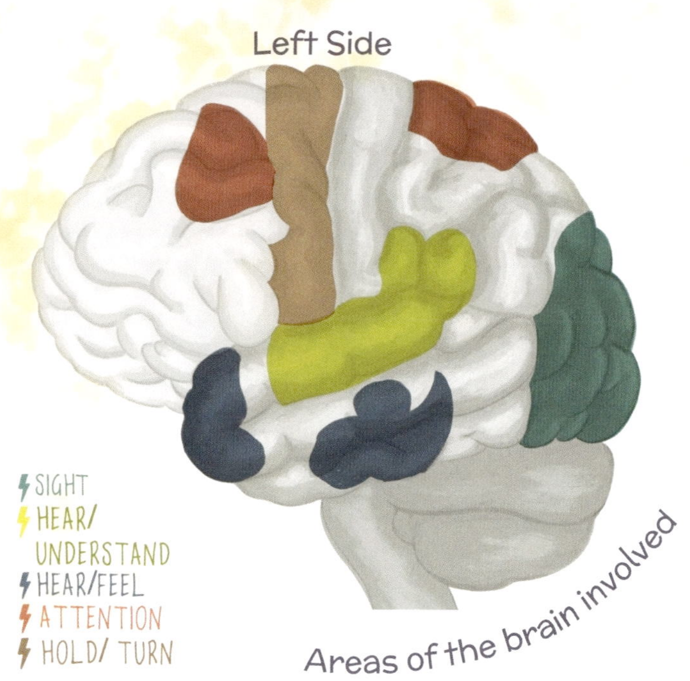

"Shared" reading together builds **interest** and fun, healthy reading **routines**.

Snuggly shared reading triggers release of **oxytocin** in the brain – the "love hormone."

Listening to stories helps children understand words (**receptive language**).

Billie helped turn pages and **listened** to books read in playful ways.
This made her feel **loved** and excited for more - at any time of day! .

Reading in fun ways encourages talking or *dialogue*. This is sometimes called **Dialogic Reading.**

It's ok to read the same books over and over. Children may memorize favorite ones and "read" them to friends!

Billie learned lots of **words** from books,
even ones read many times.
From listening to **talking**, back and forth:
pictures, questions, rhymes!

AGE
BIRTH
6 mo.
1 yr.
18 mo.
2 yrs
3 yrs
4 yrs
6 yrs
8 yrs

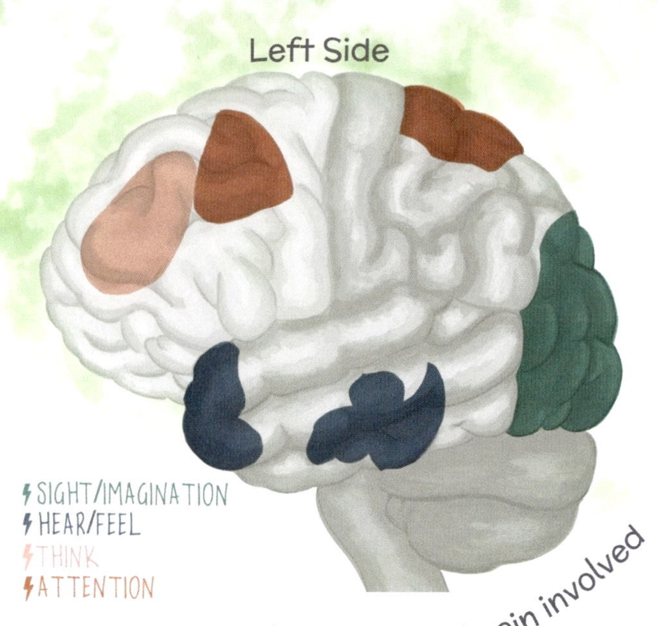

Areas of the brain involved

Reading builds **imagination**. Words and pictures are "brought to life" in the brain, including feelings and understanding.

Imagination uses the same part of the brain as seeing. That's why it's sometimes called the "**Mind's Eye.**"

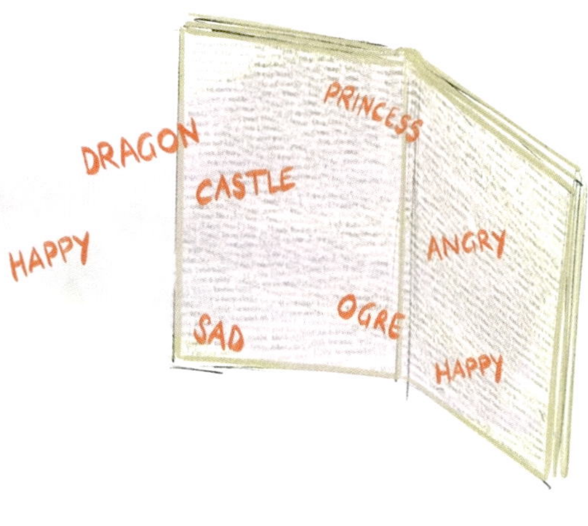

Words and pictures sparked **IMAGINATION**
that made stories seem so real.
New worlds to explore, friends to meet, feelings Billie could feel.

AGE
BIRTH
6 mo.
1 yr.
18 mo.
2 yrs
4 yrs
6 yrs
8 yrs

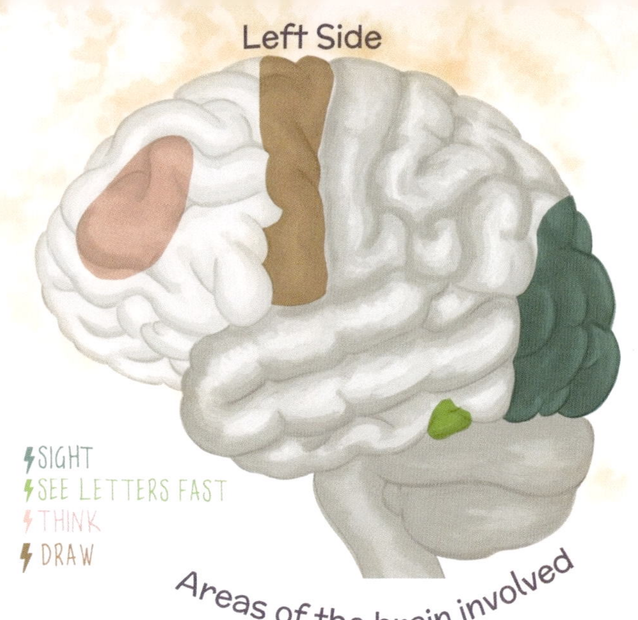

Alphabet Knowledge is learning to recognize **letters**, starting with squiggles and shapes.

The first letters a child learns are often ones in their **name** and **A–B–C**!

Ways to learn letters using multiple senses:

Billie wondered about squiggles on pages:
"What could they be?"
A voice said, "**LETTERS!**"
and one by one she learned them, *A...B...C...*

AGE
BIRTH
6 mo.
1 yr.
18 mo
2 yrs
3 yrs
4 yrs
6 yrs
8 yrs

Left Side

THINK
HEAR WORD SOUNDS
SAY/UNDERSTAND WORDS
CONNECTION

Areas of the brain involved

Phonological awareness is noticing *sounds* in words, from BIG to small.

✓ **Decoding** is taking words apart.
✓ **Blending** is putting sounds together

RHYMING
whole words or word endings..

ALLITERATION
first word sounds.

SYLLABLES
BIG word sounds.

PHONEMES
small word sounds.

BLENDING
putting word sounds together.

Sometimes children struggle hearing word sounds, like rhymes. This can be a sign of a reading problem called **dyslexia.**

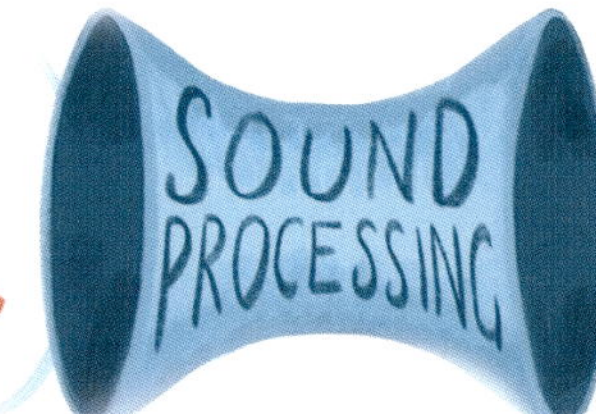

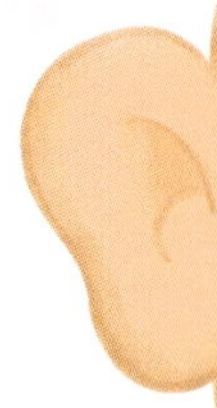

Soon, as she listened closely,
an awesome thing occurred.
Billie became aware of **rhymes**
and other sounds in words!

Letter–sound knowledge is learning the sounds that letters and letter pairs make.

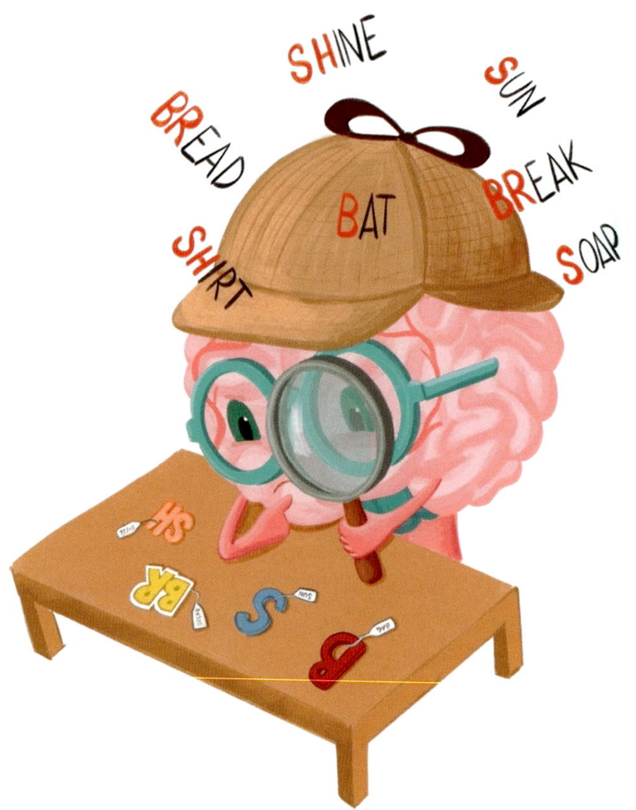

Children who struggle with letter sounds can use other senses and ways to help learn them.

Billie wondered,
"Can letters make sounds, the way that words do?"
Yes! She learned each letter could talk
and some pairs could, too!

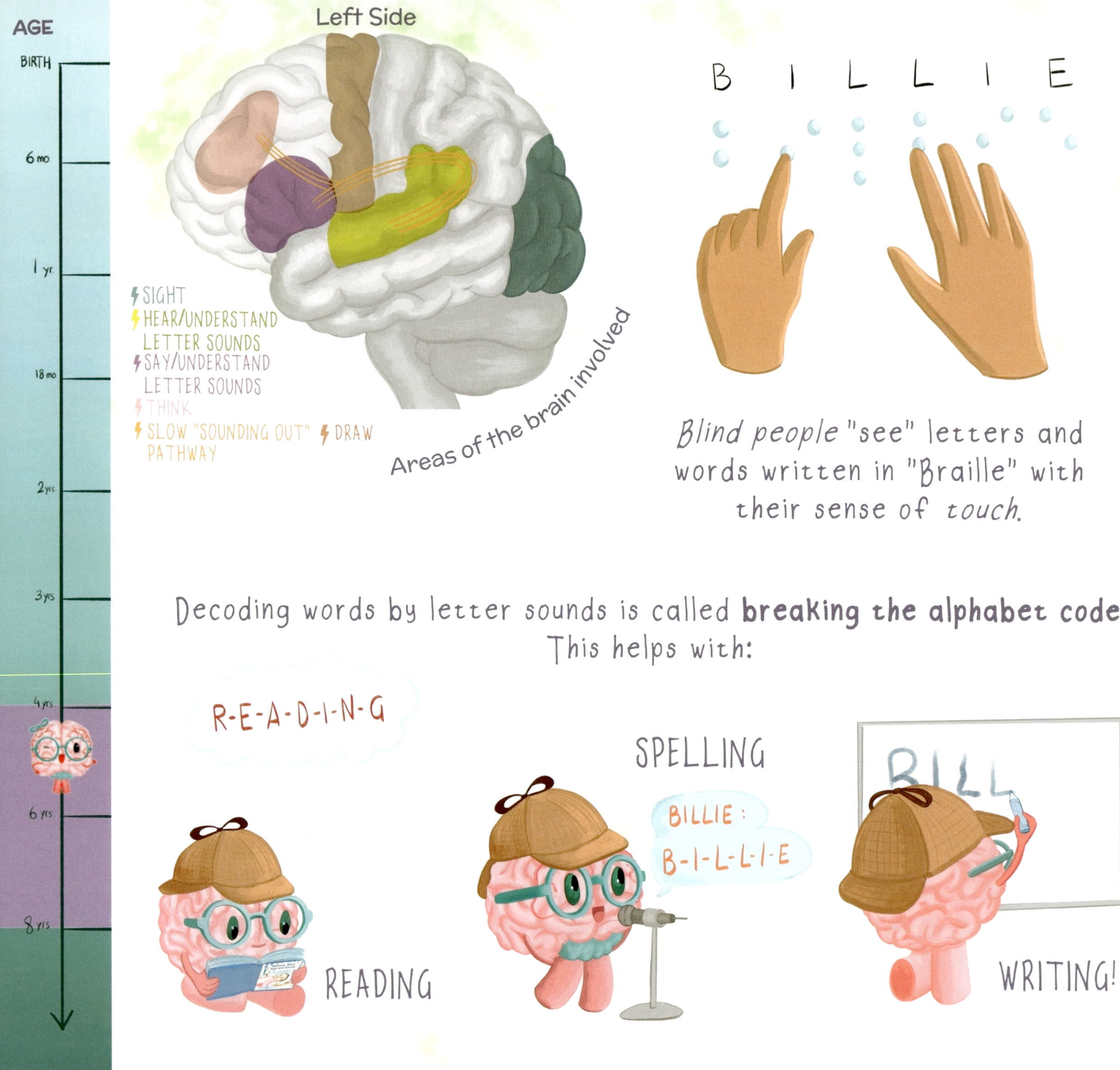

AGE
BIRTH
6 mo.
1 yr.
18 mo
2 yrs.
3 yrs
4 yrs
6 yrs
8 yrs
Left Side
SIGHT
HEAR/UNDERSTAND LETTER SOUNDS
SAY/UNDERSTAND LETTER SOUNDS
THINK
SLOW "SOUNDING OUT" PATHWAY
DRAW
Areas of the brain involved
B I L L I E
Blind people "see" letters and words written in "Braille" with their sense of touch.
Decoding words by letter sounds is called **breaking the alphabet code!**
This helps with:
R-E-A-D-I-N-G
READING
SPELLING
BILLIE:
B-I-L-L-I-E
BILL
WRITING!

If words were made of letters...
and letters all made sounds...
Billie could take words apart -
the reading code was found!

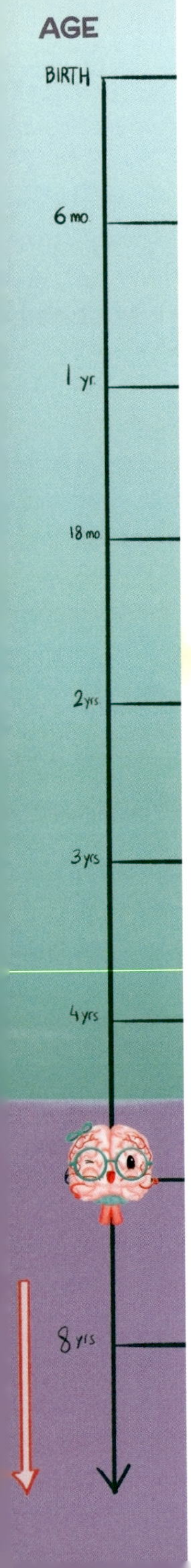

AGE

BIRTH

6 mo

1 yr

18 mo

2 yrs

3 yrs

4 yrs

8 yrs

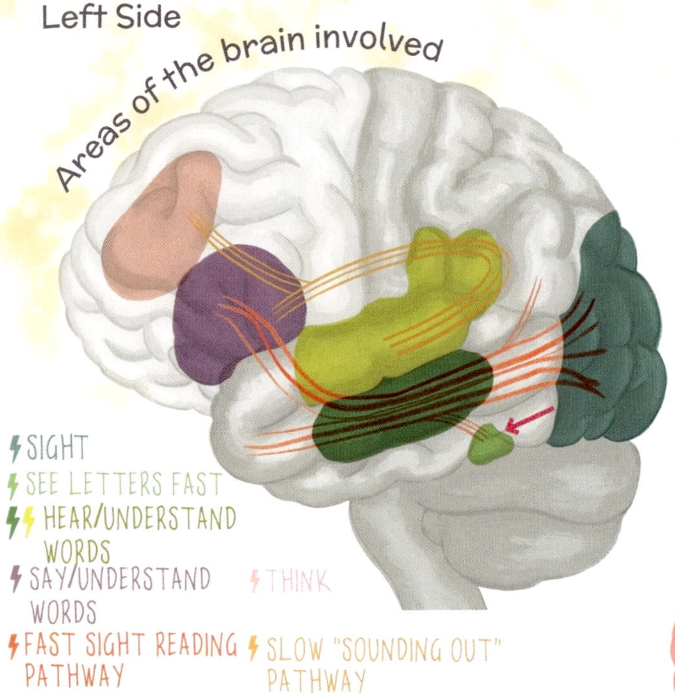

Left Side

Areas of the brain involved

SIGHT

SEE LETTERS FAST

HEAR/UNDERSTAND WORDS

SAY/UNDERSTAND WORDS

THINK

FAST SIGHT READING PATHWAY

SLOW "SOUNDING OUT" PATHWAY

Sight reading is reading words visually instead of sounding them out. It involves a part of the brain that recognizes familiar words quickly, the **Visual Word Form Area.**

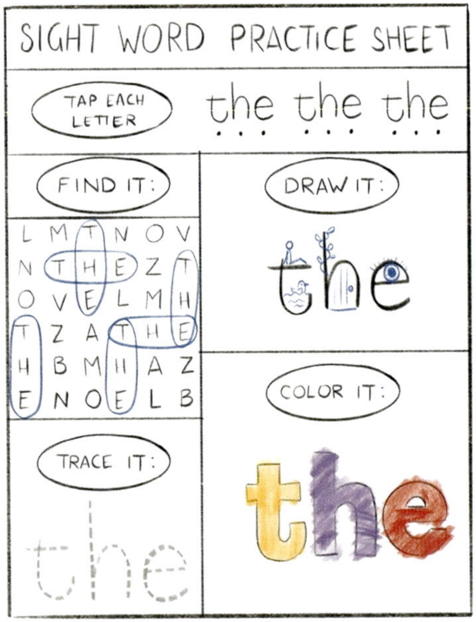

SIGHT WORD PRACTICE SHEET

TAP EACH LETTER: the the the

FIND IT:

L M T N O V
N T H E Z T
O V E L M H
T Z A T H E
H B M H A Z
E N O E L B

DRAW IT: the

COLOR IT: the

TRACE IT: the

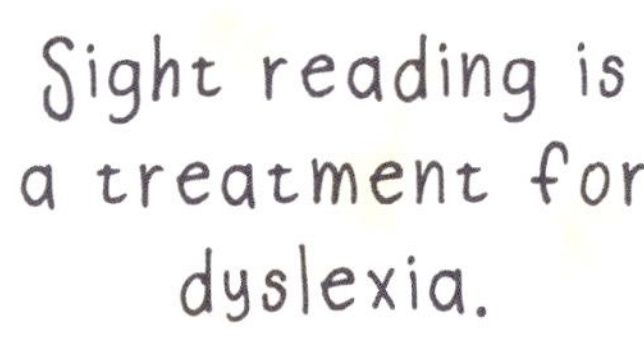

Sight reading is a treatment for dyslexia.

Billie still listened to stories
but now helped read them, too.
She **decoded** some words but read faster
by **sight** ones that she knew!

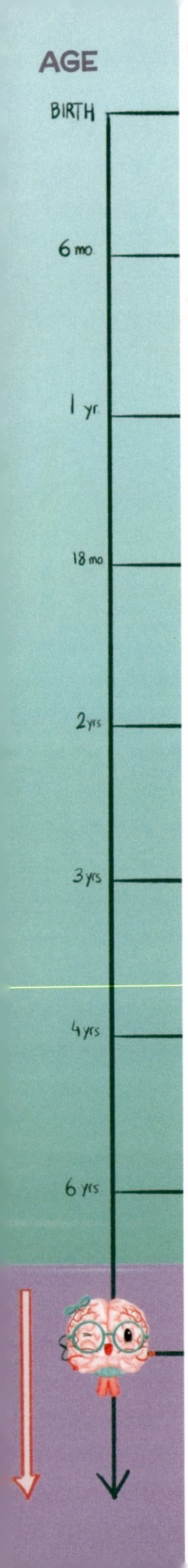

AGE

BIRTH

6 mo

1 yr

18 mo

2 yrs

3 yrs

4 yrs

6 yrs

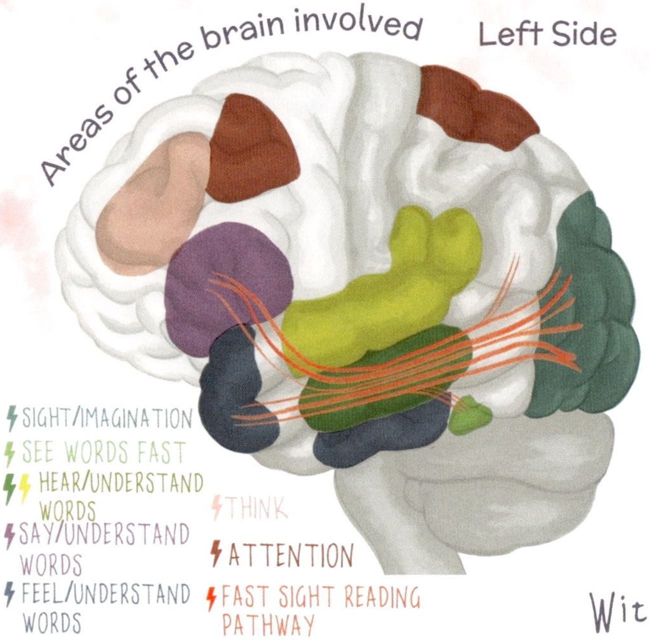

Areas of the brain involved

Left Side

SIGHT/IMAGINATION
SEE WORDS FAST
HEAR/UNDERSTAND WORDS
SAY/UNDERSTAND WORDS
FEEL/UNDERSTAND WORDS
THINK
ATTENTION
FAST SIGHT READING PATHWAY

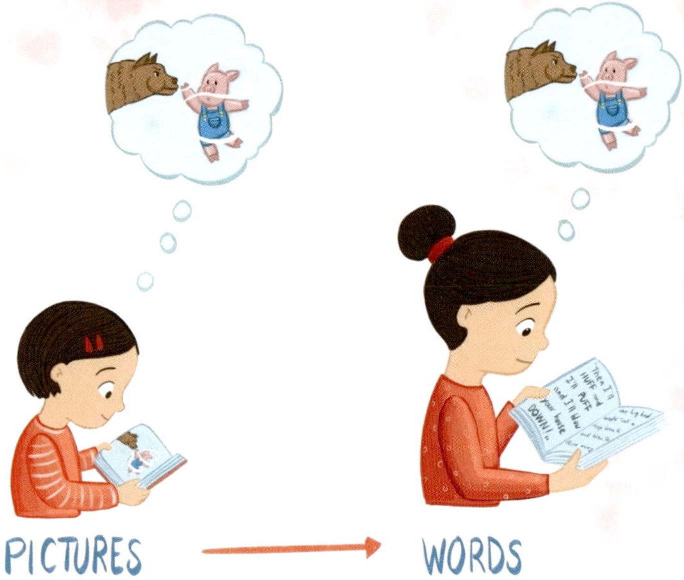

PICTURES → WORDS

With practice, "seeing" and understanding stories shifts from pictures to imagination.

Reading helps **see** and **feel** what characters are experiencing, which builds a skill called **empathy.**

MIRRORS

WINDOWS

Billie loved books with pictures
but soon she didn't need them.
Her **imagination** helped her see and feel stories
as she'd read them.

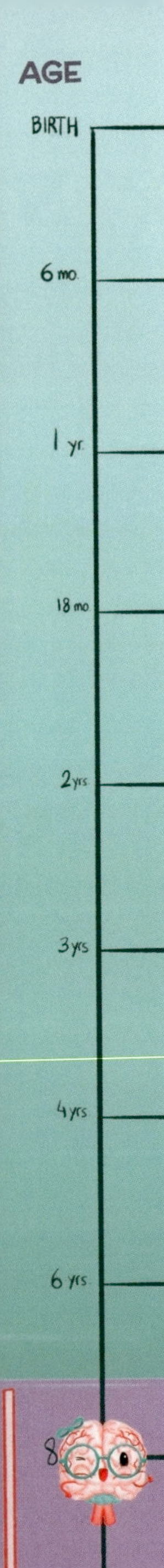

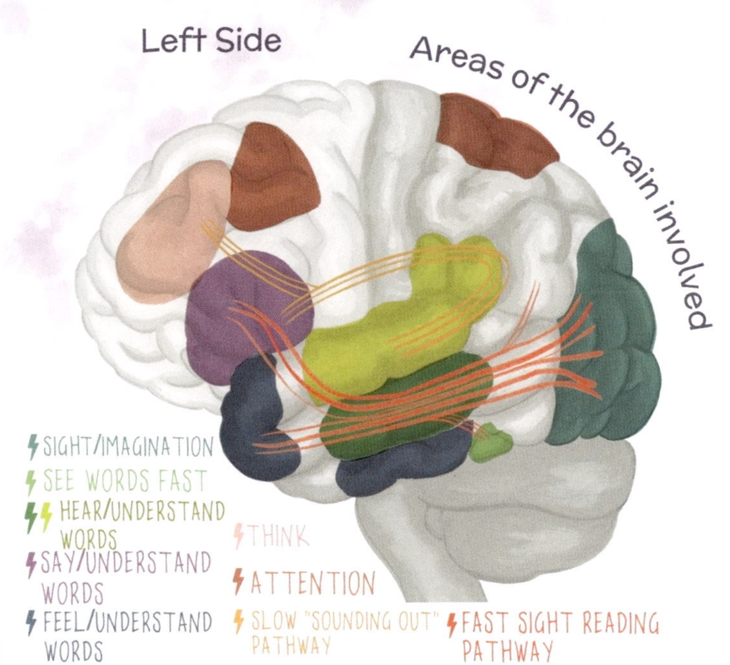

Nerve fibers (**white matter tracts**) connecting brain areas grow thicker and faster with practice. This involves a fiber coating called **myelin** (*"myelination"*).

"Neurons that fire together, wire together." — Donald Hebb.

Billie's **neurons** fired away as she read,
making connections faster and stronger.
Shaping her growing **a-Billie-ties**
to read books that were harder and longer.

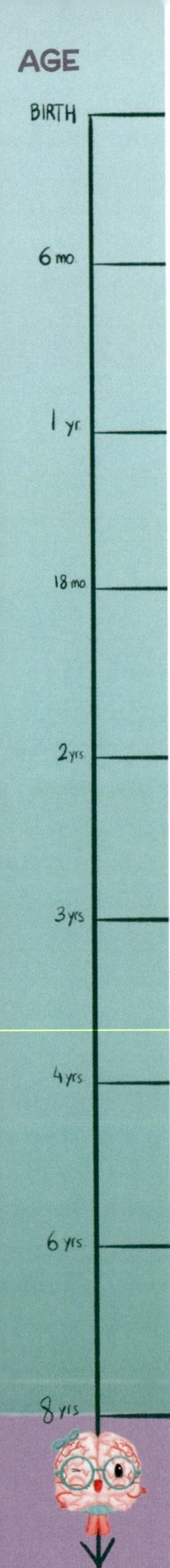

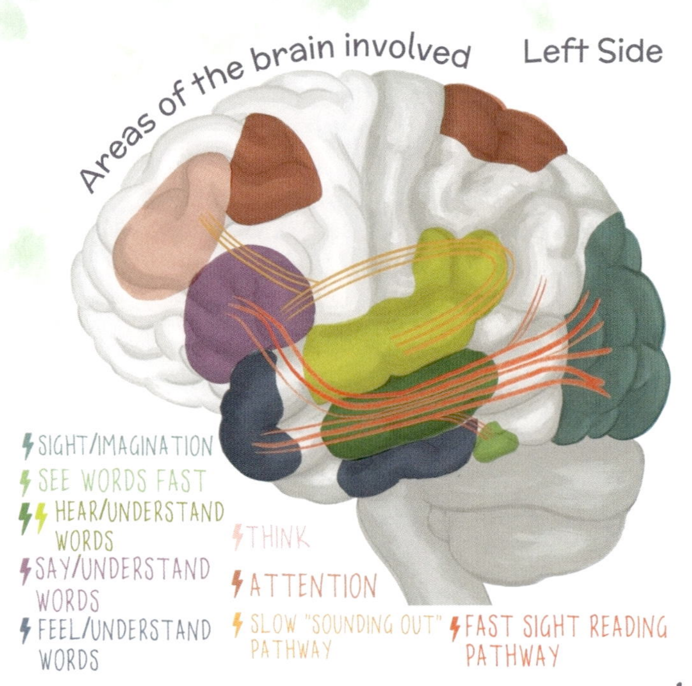

Language and Literacy Live in Left Lobes.

The reading network shifts to the left side of the brain with practice, which is more efficient.

Around 3rd grade, there is a shift in school from "learning to read" to "reading to learn." But it's always great to **read for fun!**

Fluency is reading quickly, accurately and with understanding... the final step!

And so, Billie had built it - *hooray!*
A **brain network** to help her read!
Opening new worlds and ways to learn,
happily ever after, indeed.

COMPONENT BRAIN NETWORKS
(Left Hemisphere shown)

FRONTAL-PARIETAL
(EXECUTIVE FUNCTION)
(“Thinking”)

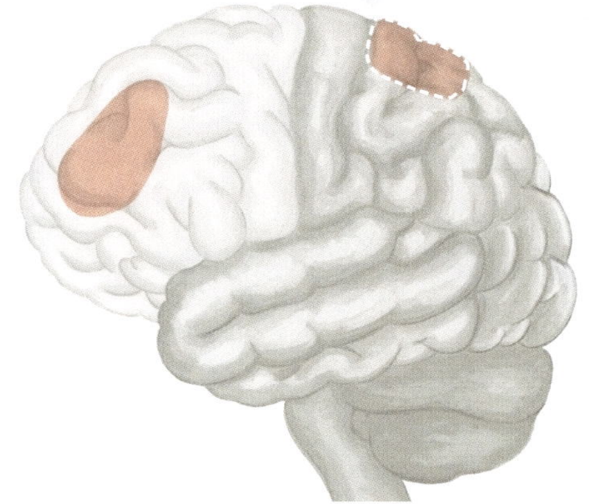

DORSAL ATTENTION
(“SPOTLIGHT” FOCUS)

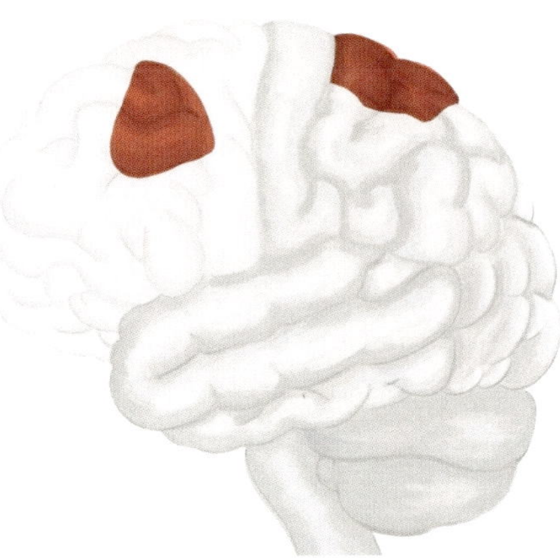

INTEGRATED

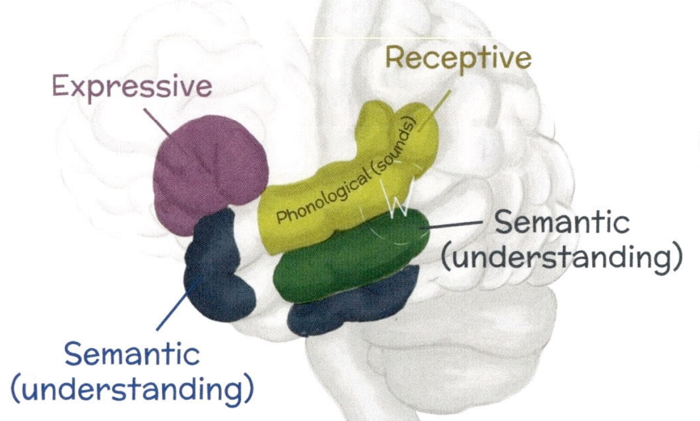

LANGUAGE

Primary Visual &
Secondary Visual
(Imagination)

Visual Word Form Area

VISUAL

Existing networks are repurposed and connected for reading.

FUNCTIONAL READING NETWORK

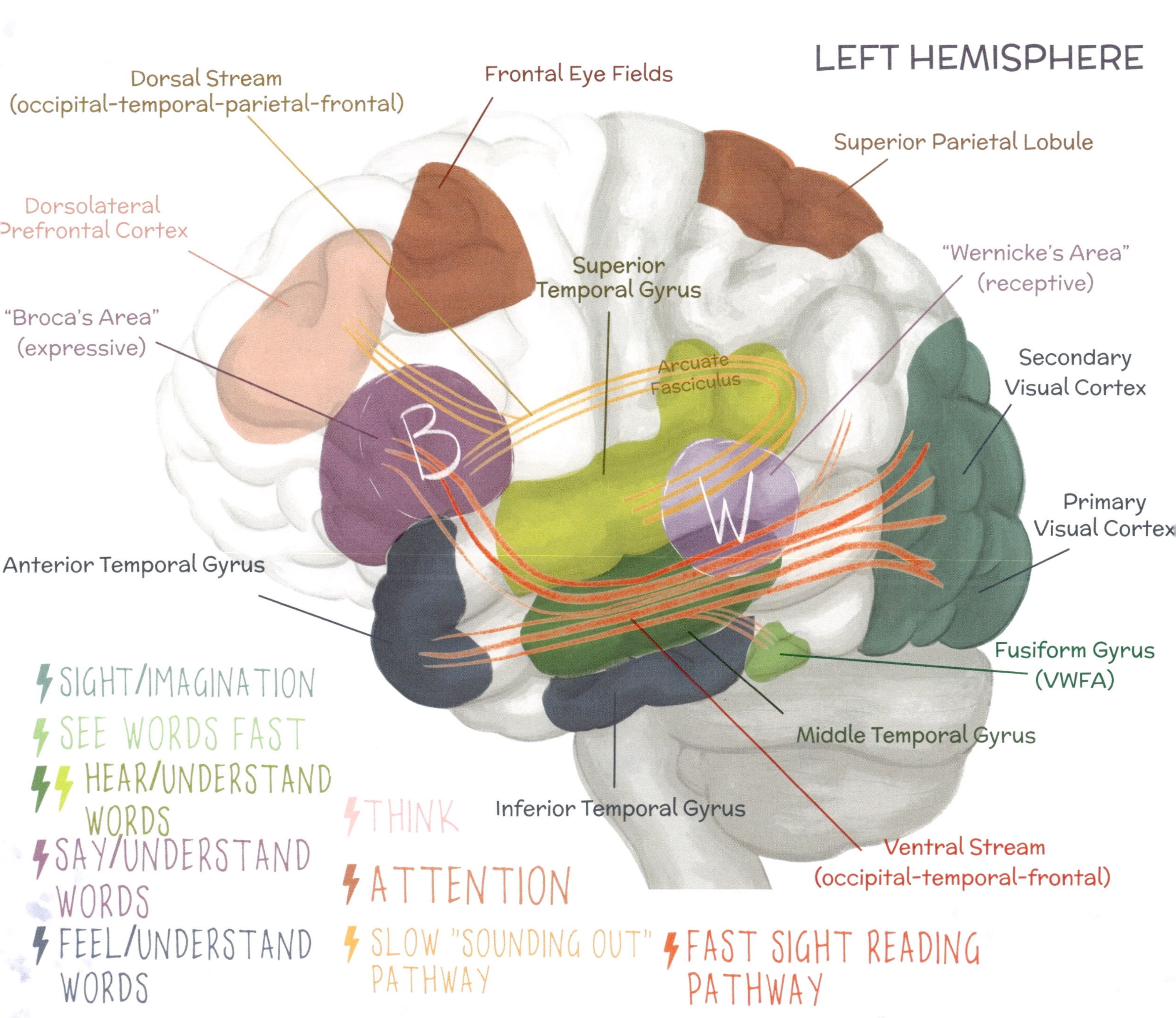

EMERGENT LITERACY GROWTH CHART

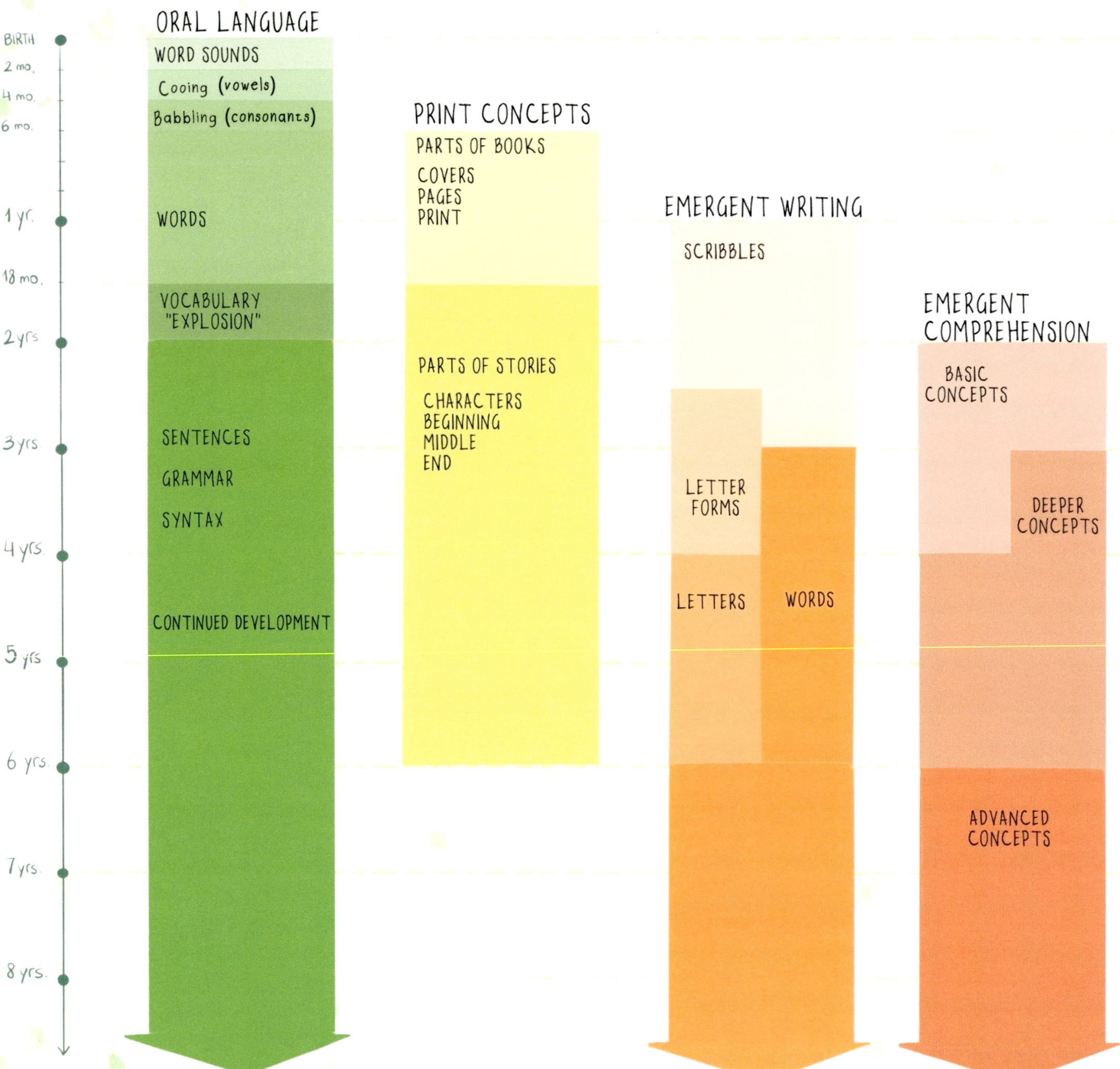

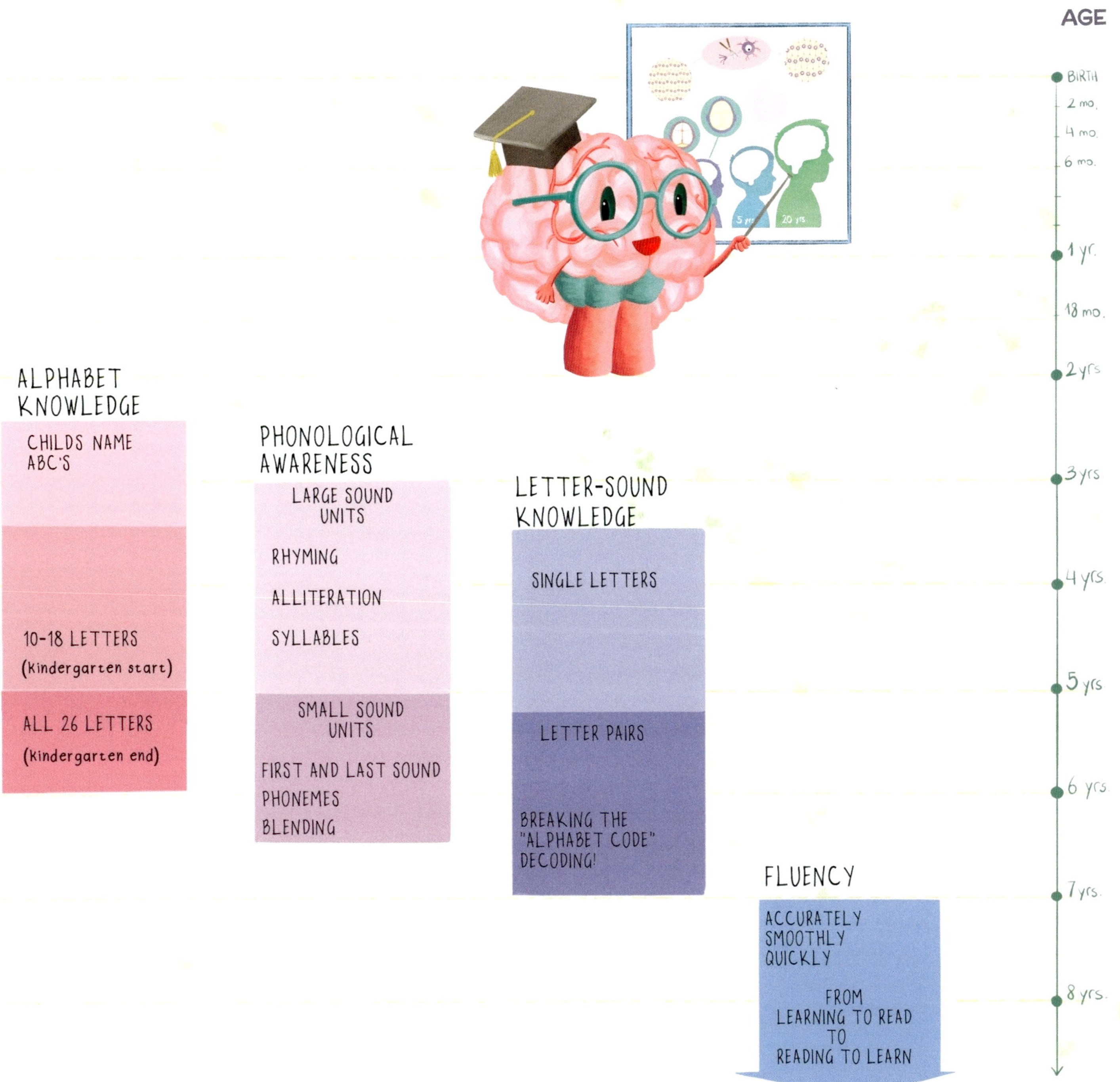
AGE
BIRTH
2 mo.
4 mo.
6 mo.
1 yr.
18 mo.
2 yrs
3 yrs
4 yrs
5 yrs
6 yrs
7 yrs.
8 yrs.
5 yrs
20 yrs
ALPHABET KNOWLEDGE
CHILDS NAME
ABC'S
10-18 LETTERS
(kindergarten start)
ALL 26 LETTERS
(kindergarten end)
PHONOLOGICAL AWARENESS
LARGE SOUND UNITS
RHYMING
ALLITERATION
SYLLABLES
SMALL SOUND UNITS
FIRST AND LAST SOUND
PHONEMES
BLENDING
LETTER-SOUND KNOWLEDGE
SINGLE LETTERS
LETTER PAIRS
BREAKING THE "ALPHABET CODE" DECODING!
FLUENCY
ACCURATELY
SMOOTHLY
QUICKLY
FROM LEARNING TO READ TO READING TO LEARN

SOURCES OF READING DIFFICULTIES

PROBLEM WITH A COMPONENT NETWORK

LANGUAGE

VISUAL ⟶ BRAILLE

ATTENTION/EXECUTIVE

PROBLEM WITH NETWORKS COMING TOGETHER

NEURODEVELOPMENT (GENETICS)

DAMAGE TO AREA OR FIBER TRACT

LESS READING PRACTICE

CONTRIBUTORS

CHRONIC AND/OR COMPLEX MEDICAL CONDITIONS

DIRECT AND INDIRECT (E.G. MISSED SCHOOL) CONTRIBUTORS
PREMATURITY (ESP < 32 WEEKS)
HEARING LOSS
ADHD
EPILEPSY
PEDIATRIC LEUKEMIAS/BRAIN TUMORS
CONGENITAL HEART DISEASE
SICKLE CELL ANEMIA
MANY MORE!

GENETICS/FAMILY HISTORY

DYSLEXIA~40–80% HERITABLE (≥ 9 GENES)
USUALLY AFFECTS PHONOLOGICAL SKILLS

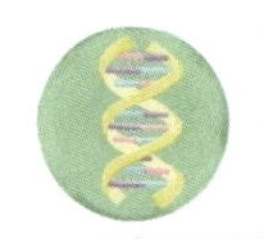

HOME LITERACY ENVIRONMENT (HLE)

ORAL LANGUAGE, PRINT CONCEPTS, INTEREST
POVERTY ⟶ HIGH RISK
EXCESSIVE SCREEN TIME (DISPLACEMENT OF READING)

HOW TO HELP

COMMON EARLY SIGNS

LANGUAGE DELAY

DIFFICULTY RHYMING LETTER SOUNDS OR SOUNDING OUT WORDS

RESISTANCE OR REFUSAL TO READ

WHEN TO ASK FOR HELP

LANGUAGE CONCERNS – AS SOON AS POSSIBLE

EARLY SKILL SCREENING – PRESCHOOL AGE

FORMAL READING TESTING – KINDERGARTEN OR OLDER

IF WORRIED, ASK YOUR DOCTOR OR TEACHER!

SOME OF READING RESOURCES AND ORGANIZATIONS

- ✓ READING ROCKETS
- ✓ REACH OUT AND READ
- ✓ DOLLY PARTON'S IMAGINATION LIBRARY
- ✓ PUBLIC LIBRARIES
- ✓ ENCOURAGE INTERACTIVE "FUN" READING:
  - → SHARE/STEP APPROACH (INFANTS)
  - → DIALOGIC READING (OLDER CHILDREN)

THE END

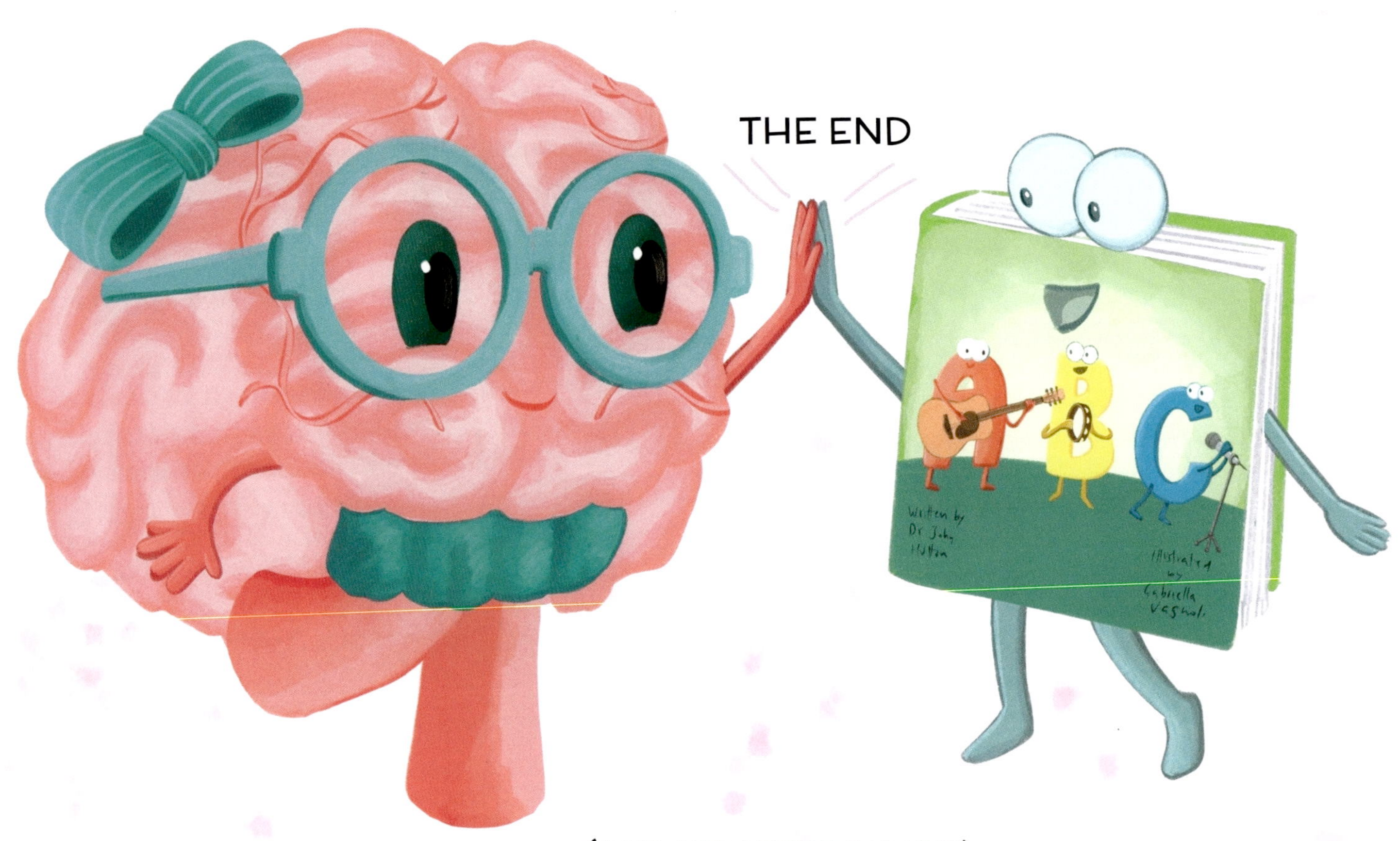

(AND NEW BEGINNINGS!)